NATIONAL GEOGRAPHIC

School Publishing

Saturn
The Ring World

PIONEER EDITION

By Lesley J. MacDonald

CONTENTS

Ring Wo

rld

○ By Lesley J. MacDonald ○

Saturn is one of the best-
known planets. Yet it is also
one of the most mysterious.
Now a spacecraft looks at
this amazing ring world.

Have you ever wondered what it would be like to fly through Saturn's rings? A **spacecraft**, or ship that flies in space, did just that.

In 2004, the Cassini spacecraft arrived near Saturn. It had spent more than seven years speeding through space. It had traveled more than 3.5 million kilometers (2.2 million miles). Now it was about to start the most dangerous part of its mission.

Through the Rings

Cassini arrived below Saturn's rings. It now had to make its way through the rings. The rings are made of bits of rock and ice. Would the spacecraft make it? Or would it crash?

Luckily, Cassini made it through the rings. The spacecraft then slowed down. It started to **orbit**, or go around Saturn.

Cassini is not the first spacecraft to go to Saturn. Voyager 2 visited Saturn in 1981. But it did not stay long. It took hundreds of photos. Then it zoomed past Saturn.

Cassini is not going anywhere else. It completed its mission to circle Saturn for four years. Now it is on an extended mission. There is a lot for scientists to learn from these missions.

Gas Giant

Saturn is a **planet**. That is a large object in space that moves around a star. Saturn is the second largest planet in our solar system. In fact, 750 Earths could fit inside it!

Saturn is made of gases. The gases may look calm. But they are not. Tornadoes and other storms whip around Saturn. Some winds blow around 1,600 kilometers (1,000 miles) an hour!

Going to Work. This art shows Cassini racing toward Saturn.

Colorful Rings.
Thousands of rings
surround Saturn.
The colors show
some of them.

Many Moons. Voyager 2 flew by Saturn in 1981. Its images appear in this NASA collage. Now Cassini is taking new, clearer photos.

Enceladus

Titan

Rhea

Saturn

Dione

Mimas

Tethys

Lord of the Rings

Saturn is best known for its rings. From far away, they look solid. Up close, things look different.

The rings are made of pieces of ice and rock covered by ice. Many chunks are the size of a grain of sand. Others are the size of a house. Some are even larger.

More than a thousand bands make up the rings. Some parts are braided together. Scientists have also found **moons** in the rings.

Meet the Moons

Saturn has more than 50 moons. Most of them are small. Many are much smaller than Earth's moon. What are some of these moons like?

Enceladus is one of the shiniest objects in the solar system. Lots of sunlight bounces off its icy surface.

Mimas has a huge crater. It covers nearly a third of the moon's surface.

Tethys has long deep trenches. It also has tiny moons. Imagine, a moon that has smaller moons!

Moon Mission. This picture shows Huygens landing on Titan's surface.

Mystery Moon

The most amazing moon of all is Titan. A thick **atmosphere**, or layer of air, covers the moon. Some scientists say there may be life on Titan.

To find out, Cassini carried a small spacecraft to Saturn. It is named Huygens. Huygens landed on Titan. Then it studied the moon's surface and atmosphere. Huygens found rivers on Titan. Perhaps Huygens will make many other discoveries in the future.

Wordwise

atmosphere: layer of air around a planet or moon

moon: natural obJ.ect that circles a planet

orbit: to go around

planet: large object that goes around a star

spacecraft: ship that flies through space

Cassini and Huygens give us great views into space. But what is the story behind them? Why do they have such odd names?

The spacecraft are named after two scientists. The two men lived long ago. They had only simple tools for studying Saturn. Yet they learned a lot about this planet.

Finding Rings

Christiaan Huygens lived in the 1600s. This was long before any spacecraft. So Huygens studied Saturn with a telescope. That is a tube with curved glass inside.

Huygens saw things that no one had seen before. He discovered the planet's rings. At first, people thought he was crazy. No one had heard of rings around a planet. But soon Saturn became known for its rings.

Ring Research. Huygens BELOW made this drawing of Saturn moving around the sun RIGHT.

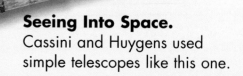

Seeing Into Space. Cassini and Huygens used simple telescopes like this one.

Name?

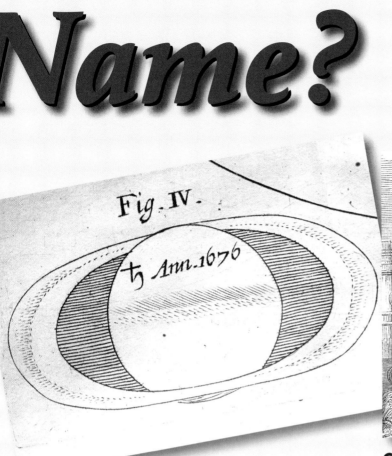

Fig. IV.

♄ Ann. 1676

Cassini's Studies. Cassini ABOVE studied Saturn. This drawing shows a gap he found in the rings LEFT.

Filling the Gap

Gian Domenico Cassini lived at about the same time as Huygens. He studied many different planets. Yet he is best known for his studies of Saturn.

Cassini was the first person to see four of Saturn's moons. He figured out that Saturn's rings are made of rock. He also found a gap in the rings. These discoveries made him famous.

Days of Discovery

Cassini and Huygens used telescopes. They also used their minds. They changed the way people thought about space.

Scientists like Cassini and Huygens made many discoveries. But there is still much to learn.

Each day, we find out more about space. Every discovery leads to new questions. Someday you might help find the answers.

Changing How

A Mountaintop View.
This building protects
a powerful telescope
on a mountaintop
in Hawaii.

Much has changed since the 1600s. Today, we have better tools. Telescopes look deeper into space. They give us amazing new views.

Some telescopes are small. They let us study stars from our backyards. Scientists use telescopes with more power. These can see far into space.

Getting a Better Look

Many large telescopes sit on tops of mountains. The sky there is clear. The telescopes have good views.

Yet the sky always makes some objects look fuzzy. So scientists have found ways to get an even better view. They send telescopes into space.

We See Space

Seeing in Space.
The Hubble Space Telescope orbits Earth. It gives clear views deep into space.

Views From Space

In space, telescopes travel above the sky. Their view is crystal clear. They show details that we cannot see from Earth.

Some telescopes orbit, or circle, Earth. Others travel deep into space. They show us faraway worlds.

Pictures of Pluto

In 2006, a spacecraft was sent to Pluto. It will be the first ever to visit this icy world.

The trip will take eight or nine years. Scientists think it will be worth the wait. The pictures may change what we know about space.

Saturn

**Take a spin at these questions
to find out what you've learned.**

1 How would you describe the planet Saturn?

2 What are Saturn's rings made of?

3 Why was the Huygens spacecraft sent to Titan?

4 Who were Cassini and Huygens?

5 How do scientists study Saturn?